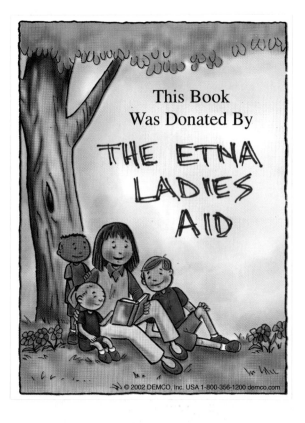

This Book
Was Donated By

THE ETNA
LADIES
AID

© 2002 DEMCO, Inc. USA 1-800-356-1200 demco.com

ANIMALS BY HABITAT

ANIMALS
OF THE
OCEANS

STEPHEN SAVAGE

RAINTREE
STECK-VAUGHN
PUBLISHERS

A Harcourt Company

Austin   New York
www.raintreesteckvaughn.com

Titles in the Animals by Habitat series
Animals of the Desert
Animals of the Grasslands
Animals of the Oceans
Animals of the Rain Forest

© **Copyright 1997, text, Steck-Vaughn Company**

Published by Raintree Steck-Vaughn Publishers, an imprint of Steck-Vaughn Company

Library of Congress Cataloging-in-Publication Data
Savage, Stephen.
Animals of the oceans / Stephen Savage.
    p.   cm.—(Animals by habitat)
    Includes bibliographical references and index.
    Summary: Describes the world's seas and oceans and the mammals, fish, reptiles, and invertebrates that live in them.
    ISBN 0-8172-4753-X
    1. Marine animals—Juvenile literature.
    2. Marine ecology—Juvenile literature.
    3. Ocean—Juvenile literature.
    [1. Marine animals. 2. Ocean.]
    I. Title. II. Series: Savage, Stephen, Animals by Habitat.
    QL122.2.S28 1997
    591.92—dc20   96-35726

Printed in Italy. Bound in the United States.
3 4 5 6 7 8 9 0 02

Habitat Maps
The habitat maps in this series show the general distribution of each animal at a glance.

Picture Acknowledgments
Heather Angel 28, 29 (bottom right); **Bruce Coleman Ltd** Jeff Foott Productions 11 (top left), George McCarthy 12, Jeff Foott Productions 13 (bottom left), 13 (top right), Jane Burton 15 (top right), Charles and Sandra Hood 16, Neil McAllister 17 (top right),Carl Reossler 18, Dr Frieder Sauer 19 (bottom left), Andrew J. Purcell 19 (top right), Carl Roessler 21 (right), Andrew J. Purcell 22, Jeffrey L. Rotman (top right), Jane Burton 25 (both), Jane Burton 27 (top left), John Murray (right), Nancy Sefton 29 (top left); **Frank Lane Picture Agency** B. Borrell 4, R. Tidman 5 (main picture), Panda photo 5 (inset), W. Wisniewski 7 (left), Panda/B. Cranston 9 (top left), D. P. Wilson 14, Ian Cartwright 17 (bottom left), W. Wisniewski 21 (top left), Silvestris/D. Fleetham (left); **NHPA** Norbert Wu 15 (bottom left); **OSF** Peter Parks 9 (bottom right), Tim de Roy 19, Howard Hall 11 (bottom right), David B. Fleetham 20, G. I. Bernard 24, Fredrik Ehrenstrom 26; **Stephen Savage** 7 (right). All artwork is supplied by Linden Artists.

Contents

Words that are printed in **bold** in the text are explained in the glossary on pages 30–31.

Introduction

The oceans form the world's largest habitat, covering 70 percent of the Earth's surface. This habitat is home to many strange, colorful and unusual animals, including whales, fish, octopuses, starfish, and crabs. Some live in the shallow, sunlit waters near land or swim in the open ocean, while others live on the seabed. In shallow waters, thousands of different types of seaweed grow toward the sunlight, producing life-giving oxygen. Some animals eat seaweeds, whereas others use them as shelter.

Apart from the sound of crashing waves on the shore, the sea was once thought to be a silent place. But beneath the surface, sound is used for communication. Whales and dolphins communicate by making a variety of chirps, whistles, and squeaks. Fish use a variety of grunts, shellfish snap their shells closed, and lobsters rasp their **antennae** against their heads to communicate with each other.

Beneath the surface of the ocean is a world teeming with wildlife. ▼

Sometimes the sea is calm with barely a ripple as sunlight dances on its surface. This can soon change, and waves may heave and roll, their tops whipped into a white foam by the wind. In the dark ocean depths it is always night. Hidden in the gloom are enormous canyons and huge, underwater mountains that would dwarf even the biggest mountain ranges found on land.

These common terns and other seabirds catch fish by diving into the sea. ▶

Human use of the oceans often conflicts with the needs of ocean wildlife. Huge fishing nets often kill turtles, dolphins, and sealions, as well as the fish they were intended to catch. Almost all the great whales have been hunted to the edge of **extinction**, and they now face death from invisible poisons. The oceans are in danger from pollution, and ocean life may be at risk.

▲ Too late, humans now realize the terrible problems that their pollution causes to wildlife.

Oceans of the World

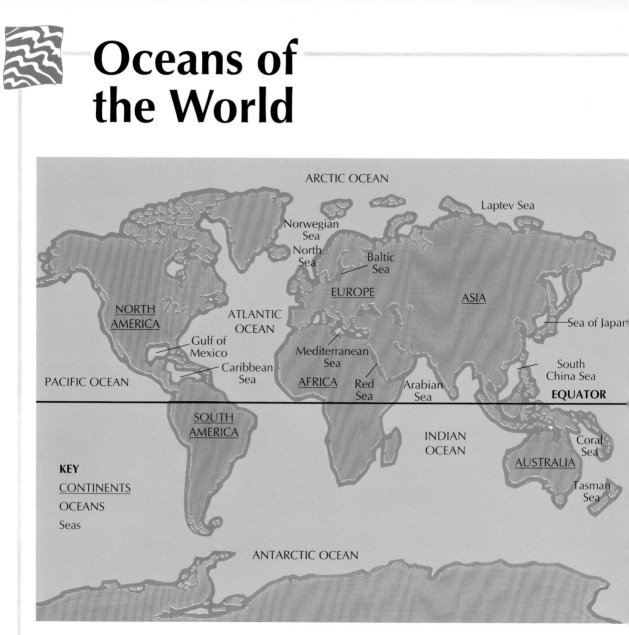

KEY

<u>CONTINENTS</u>

<u>OCEANS</u>

Seas

▲ This map shows some of the world's seas and oceans. There are many other smaller seas and bays that support a rich variety of life.

The oceans are so vast that when seen from space, the earth looks like a blue jewel. Together, these vast bodies of water surround the continents and stretch from the **tropical** waters of the equator to the frozen **polar** areas. Ocean water is constantly moving, affected by tides and currents. Some currents are warm and some are cold, but both affect the climate of the oceans.

The ocean habitat is divided into smaller habitats, called oceans and seas, which differ in size and temperature. The main oceans are the Pacific Ocean, the Atlantic Ocean, the Indian Ocean, and the Arctic Ocean. Smaller bodies of water, the seas, include the Caribbean Sea, the Mediterranean Sea, and the Sea of Japan. Many sea animals prefer to live in warm water, while others prefer it to be cold. The temperature of the water determines the species of animals that live there.

Breathtaking scenery sculpted by nature ▼

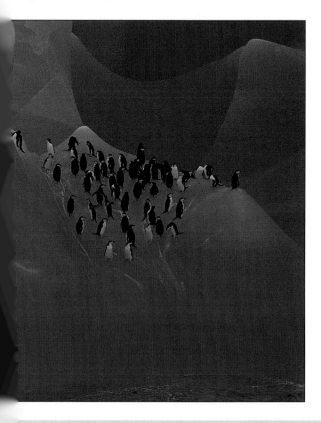

The sea has a great effect on the world's weather. Water **evaporates** from the sea to form clouds that travel over land and fall as rain. Much of this water eventually runs back to the sea.

▲ Many unusual animals live in tide pools, such as this one on a Canary Island beach formed by volcanic rock.

The sea's salty taste comes from sodium chloride (common salt) that has dissolved into the water. However, the sea contains many other dissolved minerals washed from the land, including tiny amounts of gold. Gases such as oxygen are also dissolved in sea water and sea animals extract the oxygen from water in order to breathe.

The Ocean Habitat

T he ocean habitat provides many different conditions that support a wide variety of plants and animals. This habitat stretches from seashore to dark ocean depths and across the world from the warm, tropical oceans to the icy poles. However, sunlight, water temperature, and water depth are invisible barriers that many animals cannot cross.

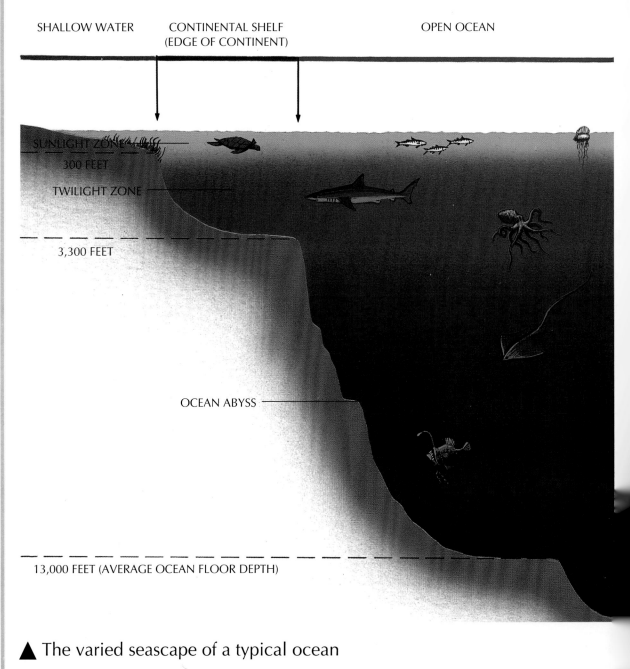

SHALLOW WATER   CONTINENTAL SHELF
(EDGE OF CONTINENT)   OPEN OCEAN

SUNLIGHT ZONE
300 FEET

TWILIGHT ZONE

3,300 FEET

OCEAN ABYSS

13,000 FEET (AVERAGE OCEAN FLOOR DEPTH)

▲ The varied seascape of a typical ocean

▲ Large kelp seaweeds are very tall and create vast, underwater forests.

Seaweeds of many shapes and sizes grow in coastal waters, and like land plants they need sunlight to survive. As the seabed gently slopes deeper, away from the sunlight, fewer seaweeds are found. Many sea creatures grow on rocks and artificial structures. Where there is no shelter from danger, some types of fish and other animals bury themselves in the sand.

Coral reefs are made by tiny coral animals. It takes thousands of years for a large reef to be built. Coral reefs provide shelter for many sea creatures and hunting grounds for larger fish. Reefs may also act as natural sea defenses, protecting islands from the full force of tropical storms. Shipwrecks also provide homes for wildlife—sponges and sea anemones grow on the outside, while fish shelter within.

Sediments and **nutrients** sink to the ocean floor and are brought to the surface by storms and ocean currents. This is food for plankton, microscopic animals, and plants, on which the ocean **food chains** depend. Humans have yet to explore most of the ocean depths. This is a strange world where the animals do not resemble animals found on land.

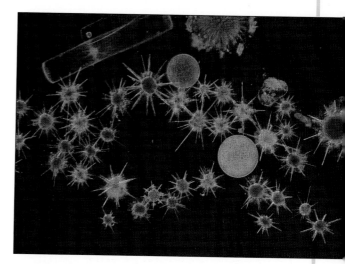

▲ Plankton are eaten by small animals, which are themselves eaten by larger animals.

Mammals

There are about eighty different types of whales, dolphins, and porpoises living in the world's oceans. These mammals have smooth bodies to help them glide through the water, and their limbs have become flippers, used for steering. Because they have lungs, they have to come to the surface of the water to breathe air.

The blue whale is so huge that its largest blood vessel is big enough for a human to crawl through. ▼

Blue Whale

At 108 feet, the blue whale is the largest animal that has ever lived. It weighs the same as forty adult elephants and has a heart the size of a small car. Despite its great size, the blue whale feeds on tiny, shrimplike animals called krill, which swim in large schools in polar seas. The whale strains large numbers of krill from the water using hairy **baleen** plates, which hang from the roof of its mouth.

▲ The Bottlenosed dolphin swims by moving its tail up and down, rather than side to side like a fish.

Bottlenosed dolphin

The bottlenosed dolphin is the type of dolphin most frequently seen by humans, because it often swims close to the shore. Dolphins use **echolocation** to find their way around and for hunting fish, which they catch with their sharp, pointed teeth. They surface to breathe, but can only stay underwater for seven minutes. Bottlenosed dolphins live in groups, and when a dolphin calf is born, the other females in the group may help look after it.

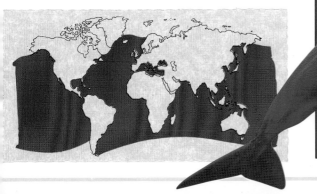

Sperm Whale

Some whales have teeth, and the sperm whale is the largest of this type, reaching 60 feet in length. They live in groups of up to thirty females and calves and are protected by a single large male. Sperm whales are able to dive down to a mile beneath the sea and can stay underwater for more than an hour. They hunt for giant squid as large as 40 feet.

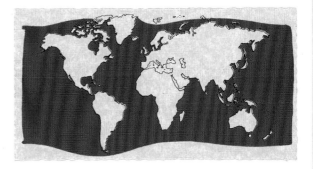

Sperm whales find their prey using echolocation by producing sounds and listening for the echo of their prey. ▼

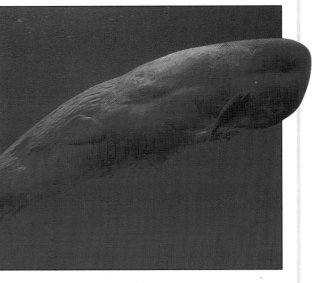

MAMMALS

Although whales and dolphins spend their whole lives in water, some types of sea mammals spend time on land. They belong to a group of mammals called pinnipeds, which means "fin-footed." Like whales and dolphins, pinnipeds' bodies are covered in **blubber** to protect them against the cold water. However, they also have hair to help keep them warm on land.

Common seal

The common seal has a plump, rounded body and large eyes that provide good underwater vision. It uses large back flippers for swimming and pulls itself onto land using its shorter front flippers. Seals often climb out onto sandbanks and rocky ledges and move by shuffling along on their fronts. A female seal gives birth to a single pup, which she feeds on milk for six weeks.

▲ The common seal has whiskers on its face that can detect the smallest vibrations of moving fish, so even a blind seal can catch its prey.

California sealion

Unlike the common seal, the California sealion uses its large front flippers for swimming. It can live to a depth of 900 feet and swim at 25 mph. Large numbers of California sealions congregate each year on special breeding beaches, where the females give birth to their pups. The female will feed her pup for about six months, by which time it will be playing with other young sealions.

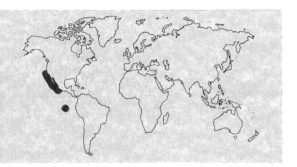

A California sealion can turn its back flippers underneath its body and walk or run on all four flippers. ▼

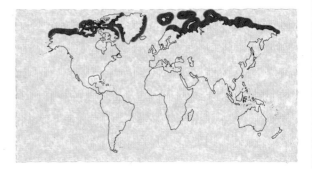

▲ The tusks of a male walrus may be up to 3 feet long.

Pacific walrus

Walruses appear to enjoy one another's company, because they form groups on beaches and ice floes. They also swim together in herds when the sea ice advances and retreats each year. Their large bodies protect them against the cold, and at 12 feet long walruses are the second largest of the pinnipeds. Walruses have large tusks to stir up the seabed in search of cockles and clams. Their short whiskers help them detect food, which includes fish.

Fish

Ocean fish come in a wide variety of shapes, colors, and sizes, including flatfish, pipefish, pufferfish, and brightly colored angelfish. The bodies of most fish are covered in scales, and they breathe oxygen from the water using gills.

The attractively marked mackerel is closely related to the giant tuna fish. ▼

Mackerel

There is no place to hide in the open ocean, so many fish, such as mackerel, live together in large schools. Living in a school means that there are many pairs of eyes to discover food or spot **predators**. Mackerel feed on small fish and swimming **crustaceans**, but are in turn then eaten by larger fish, sharks, sea mammals, and humans. A single female mackerel may lay half a million eggs, although many will be eaten by other animals.

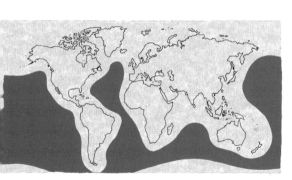

Deepsea anglerfish

In the depths of the oceans live many unusual fish, including the bizarre anglerfish. It has a rounded body, a mouthful of sharp teeth, and a living fishing rod on its head. The end of the fleshy fishing rod is luminous and resembles a tasty worm. As an attracted fish swims closer to investigate, the anglerfish moves the lure in front of its mouth. The anglerfish is a poor swimmer and often rests on the seabed.

The anglerfish quickly opens its mouth, and its prey is sucked in and swallowed whole.

▲ When threatened, the porcupine pufferfish inflates its body like a balloon and the spines stick out to protect it.

Porcupine pufferfish

The porcupine pufferfish is one of many types of pufferfish living in tropical seas. This type of pufferfish is covered with spines that lie flat against its body. The pufferfish's teeth are fused to make a tough beak, which breaks the shells of hard-bodied creatures. It feeds on a wide range of animals, including crabs, sea urchins, and sea snails. The porcupine pufferfish lives on coral reefs, in sandy bays, and **mangrove swamps**, where it grows to 20 in.

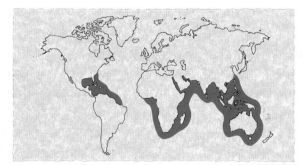

Some fish swim together for protection, while others disguise their bodies or hide from view. They have good reason to be cautious, because the seas are patrolled by predatory fish of all sizes. These predators' mouths are equipped with sharp teeth for holding a slippery meal, which is often swallowed whole.

Lionfish

The lionfish is both beautiful and deadly, and its brightly colored body and fins warn other sea creatures of its deadly nature. Hidden among the delicate fins are large, poisonous spines, which provide a very effective defense against other predators. Lionfish often live on coral reefs, where they eat a variety of fish and crustaceans. The lionfish sometimes uses its large fins to force its victim in front of its mouth.

◀ The lionfish is also known as the turkeyfish, dragonfish, zebrafish, and red firefish.

Moray eel

The moray eel has a long, snake-like body with a large head and a mouth filled with needle-sharp teeth. It is a **nocturnal** hunter of fish and octopus and spends the day hidden in rocky crevices. Large **prey** may be grabbed by the eel's jaws and a mouthful removed by twisting its body around. The moray eel is not usually harmful to humans and will retreat into a crevice if approached, but it may attack if it is provoked.

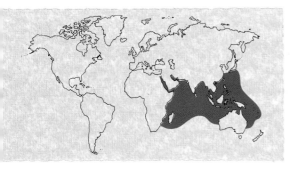

Moray eels can sometimes be encouraged to leave their lairs and may be carefully hand-fed by a diver.

▲ The spotted grouper can rapidly change the color of its body to match its surroundings.

Spotted grouper

The spotted grouper is a master of stealth and ambush. It can change color and lurks in wait for smaller fish that swim too close. When a fish is in range, the grouper lunges forward, opening its mouth and sucking the fish inside. Like some other fish, the grouper spends its first five years as a female and can change into a male when it is seven years old.

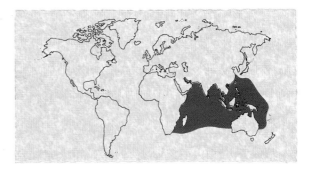

Sharks and rays are different from most fish because their skeletons are made of **cartilage** instead of bone. Sharks come in a variety of shapes and sizes, including the 60-foot whale shark, the world's largest fish. Rays are also varied. Some have a sting or catch their prey using electricity, while the 22-foot manta ray steers plankton into its mouth using extended fins at the side of its head.

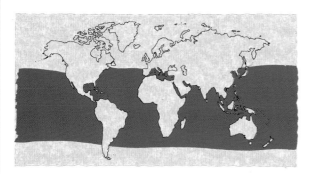

The 21-foot great white shark has strong jaws and serrated teeth, making it a powerful predator. ▼

Great white shark

The great white shark is the most dangerous predator in all the oceans. Out of the 344 types of shark, it is one of only a handful of sharks that actually attack humans. When humans are attacked, the shark probably mistakes the victim for its natural prey, which includes fish and seals. The great white shark has an exceptional sense of smell and can detect the blood of injured prey from 1,300 feet away.

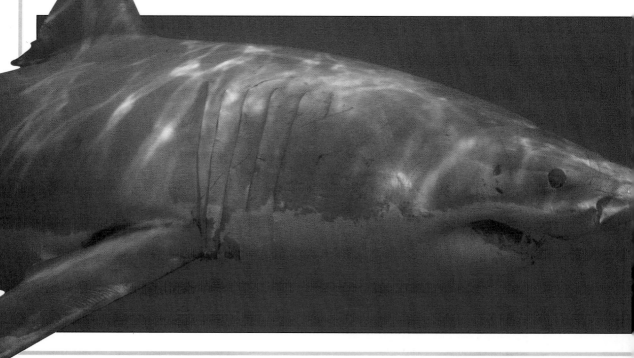

Lesser spotted dogfish

At 30 inches, the lesser spotted dogfish is one of the smallest types of shark. Unlike the great white shark, the dogfish often rests on the seabed, where its spotted skin acts as **camouflage**. It feeds on bottom-living animals, such as crabs, shellfish, flatfish, and other small fish. The dogfish lays its eggs in protective capsules, which are attached to seaweed. Each capsule contains one egg, which will hatch after nine months.

Unlike the great white shark, which has to swim to stay alive, the dogfish can rest on the seabed. ▼

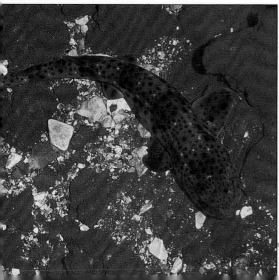

▲ A **spiracle** behind each eye allows a resting stingray to breathe without filling its mouth and gills with sand.

Blue-spotted stingray

The attractive blue-spotted stingray is one of many types of ray. It is named after the poisonous spine on its tail, which is used as a defense against predators. A stingray glides gracefully over the seabed by rippling its large, winglike fins. Like the great white shark, the female stingray gives birth to live young that hatch from eggs inside her body. The stingray's mouth on the underside of its head is filled with teeth for crushing shells.

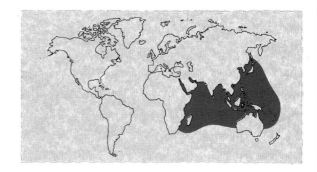

Reptiles

A few types of **reptiles** such as marine iguanas, sea turtles, and sea snakes live in the ocean. Marine iguanas spend much of their time on land, while sea turtles only return to land to lay their eggs. The banded sea snake spends all its life in the sea.

Green turtle

At 5 feet, the green turtle is one of the largest of the seven types of sea turtles. It spends most of its life at sea, and only the female returns to land to lay her eggs. She hauls herself up the beach and buries her eggs in a hole that she digs with her front flippers.

Three months later, the young turtles hatch and have to avoid many predators as they move down to the safety of the sea. Young green turtles feed on small crustaceans, fish, and jellyfish, while the adult turtles eat marine plants.

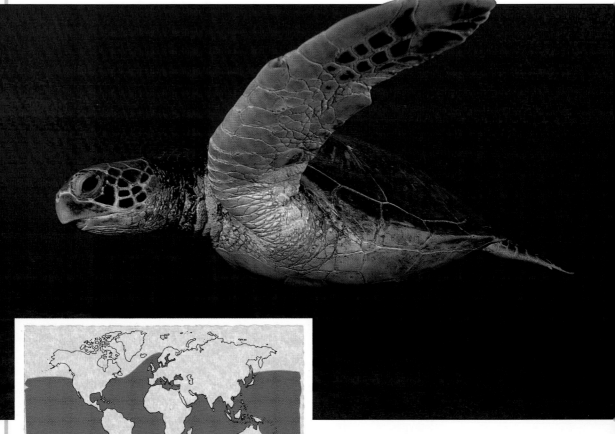

▲ Adult turtles can stay submerged for two hours.

Sea snake

The banded sea snake is fully adapted to life in the sea and lives in the coastal waters of tropical seas. It grows to 6.5 feet and has a flattened body and a paddle-shaped tail for swimming. The sea snake may lie at the surface of the sea or on the seabed waiting for prey. The female gives birth to live young that hatch from eggs inside her body.

▲ Marine iguanas warm their bodies in the sunlight before entering water.

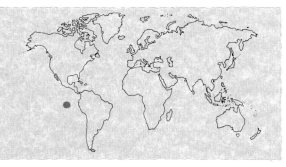

Marine iguana

The marine iguana, a type of lizard, lives on the rocky shores of the Galápagos Islands. This 3-foot-long lizard moves slowly and usually lives in groups. The shores in this region are **lava rock**, with no vegetation. The iguana feeds on seaweed beneath the waves. The iguana is a good swimmer, but it swallows stones to weigh it down in the water. The marine iguana **excretes** unwanted salt through its tear glands.

▲ A sea snake can hold its breath for two hours and has a poisonous bite that it uses to catch fish.

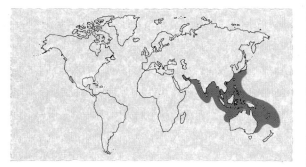

Invertebrates

Sea **mollusks** are a varied group of animals that are usually protected by a hard and often beautiful shell. Many have two halves to their shells, while others have single shells that they carry on their backs. The octopus is also a mollusk, but its walking foot is divided into eight **tentacles**.

Giant clam

Some types of mollusks are known as bivalves because they have two halves (valves) to their shells. The giant clam is the largest, reaching 18 ft. and weighing up to 395 lbs.

The clam spends its adult life settled on the seabed and feeds on plankton. The large shadow of a predator will cause the shell to close and protect the clam.

▲ The giant clam filters plankton from the seawater that it pumps in and out of its body.

Cone shell

The cone shell is an example of a mollusk that carries its single shell on its back. Many sea snails feed on algae or **carrion**, but some sea snails, such as the cone shell, are predators. The cone shell has a long feeler armed with poisonous darts that it uses to paralyze small fish and worms. The cone shell is the only snail that is poisonous enough to kill a human.

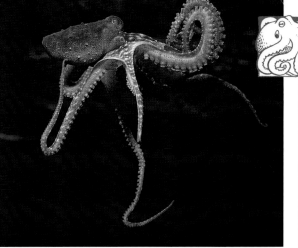

▲ An octopus's tentacles are armed with suckers for grasping a variety of prey.

▲ The beautiful but deadly cone shell travels along the seabed on its large, fleshy foot.

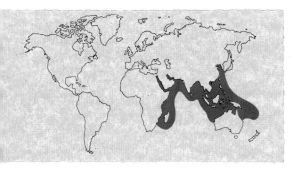

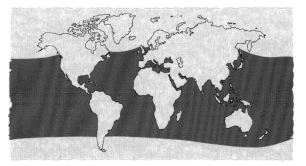

Octopus

The octopus has no need of a protective shell because it can change color and blend in with its surroundings. An octopus has a sharp beak that can break open the shells of crabs, lobsters, and bivalve mollusks. It can even squirt an inky substance to confuse a predator or large prey. The octopus rests inside its rocky lair, and the entrance is marked by a pile of empty shells. An octopus moves around by swimming, crawling, or sometimes by using jet propulsion to force water from its body.

INVERTEBRATES

C rabs, lobsters, and shrimps are all types of crustaceans. They have soft bodies that are protected by armored shells with jointed limbs to allow movement. As they grow, a crustacean sheds its hard shell, which is then replaced by a larger one from underneath.

Shore crab

The shore crab is one of many types of crabs that live in the sea. It eats small sea animals, carrion, and seaweeds. If the crab loses a limb, it can grow a new one. Each time it sheds its shell, the new limb will have grown a bit bigger. The female carries her eggs underneath her body. The newly hatched young crabs start life by swimming as part of the plankton.

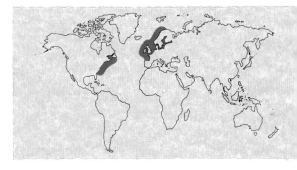

The shore crab has two large claws that are used when feeding or defending itself against predators. ▼

▲ The larger of the lobster's two claws is used for crushing shells, while the other one is used to pick up food.

Cleaner shrimp

The brightly colored cleaner shrimp leads a very different life from most other shrimps. This shrimp actually eats parasites and dead skin from larger fish that would normally gobble up a small shrimp. The fish go to special places on coral reefs known as "cleaning stations." They allow the cleaner shrimp to climb over their bodies and even their mouths and gills.

▲ The shrimps remove irritating dead skin and parasites that fish, such as this orange-spotted cod, cannot reach themselves.

European lobster

The lobster is one of the largest of the crustaceans, reaching 24 inches in length. It has **compound eyes** like an insect, but can also explore its world by smell, taste, and touch, using antennae. A lobster can usually walk on its jointed legs, but can swim backward when alarmed. Its main enemies are the octopus and humans. Lobsters, crabs, and shrimps breathe by using gills similar to those of a fish.

Starfish, sea urchins, and sea cucumbers are all types of **echinoderms**. Most produce larvae, which live as part of the plankton before they become adults. A typical starfish has five arms, but the crown-of-thorns starfish has twenty arms, and sunstars may have up to fifty.

Common starfish

The main parts of a common starfish are its arms, of which it has five. Each arm is attached to the small body, which has a mouth on the underside. If a starfish loses a leg, it can grow a new one. This predator feeds mainly on bivalve mollusks such as mussels. The starfish wraps its arms around the mussel, holding it with its tubed feet, which have suckers on the tips. The starfish pulls open the shell just enough to digest the living mussel.

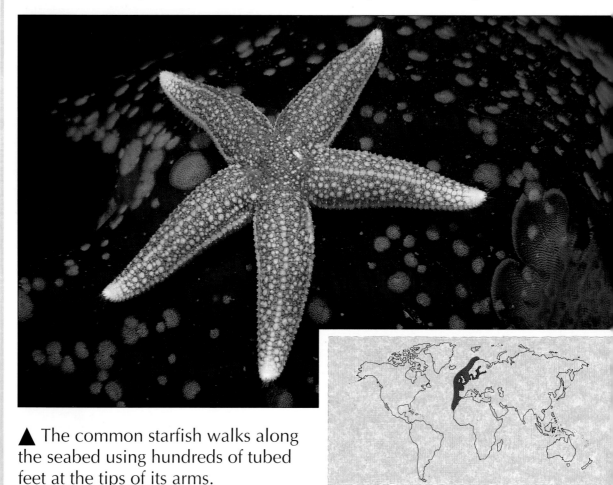

▲ The common starfish walks along the seabed using hundreds of tubed feet at the tips of its arms.

Sea cucumber

The large, sacklike body of the sea cucumber resembles a big sausage. Surrounding its mouth are several short tentacles that collect small pieces of food. Like other echinoderms, the sea cucumber cannot see and wanders along the seabed on its tubed feet in search of food. If attacked, the sea cucumber can send out a mass of sticky threads to entangle its attacker, then slowly creeps away.

Sea cucumbers are probably among the most unusual animals that live in the sea. ▼

▲ Sea urchins walk using tubed feet on the underside of their shells.

Sea urchin

The rounded body of the sea urchin is covered by a thin layer of shell. Attached to the shell are hundreds of sharp spines that help protect the urchin from being eaten. Tubed feet between the body spines may be used to carry small pieces of seaweed to disguise their bodies. The sea urchin's mouth is on the underside, and it rasps algae from rocks using five special teeth.

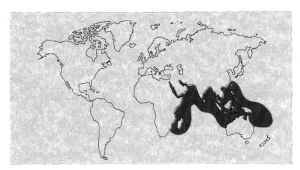

Sea anemones, corals, and jellyfish have soft bodies and stinging tentacles to catch their prey. There are many different types of these animals living in both tropical and **temperate** seas. Some types of sea anemones live in tide pools and resemble flowers.

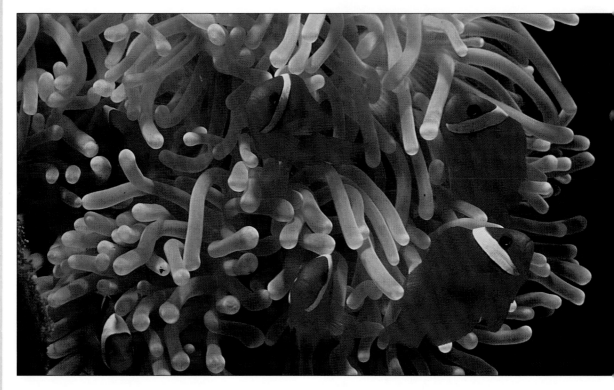

▲ These red clownfish are protected within the anemone's tentacles. In return, the fish will chase off predators trying to attack the anemone.

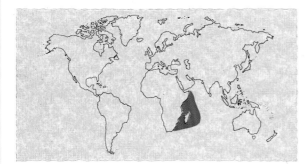

Sea anemone

A sea anemone spends its life attached to rocks by a strong disc of muscles. It has a sacklike body and a ring of tentacles with a simple mouth in the center. The tentacles are armed with stinging cells, which the anemone uses to catch small fish. A few types of sea anemone have clownfish living safely in their tentacles. The clownfish's body is covered in a special slime, which protects it against the anemone's sting.

▲ This staghorn coral is named after the antlers of a male deer (stag), which it resembles.

Staghorn coral

The staghorn coral is just one of many types of coral that grow together to form huge reefs. Each staghorn shape is made by a colony of single coral animals called **polyps**, which resemble tiny sea anemones. Tiny algae live inside the bodies of the polyps and also benefit from the protection of the stony coral branches. The polyp catches food with its tentacles, but also receives food from the algae. Most corals live in tropical seas, preferring warm, clear, sunlit water.

Lion's mane jellyfish

Unlike the sea anemone and coral, lion's mane jellyfish swim through the water by pulsating their large, bell-shaped bodies. They swim upside down with their tentacles hanging below. The jellyfish's tentacles are used for stinging and catching prey. Jellyfish are not strong swimmers and often drift along with the ocean currents. This means they are sometimes washed ashore. Do not touch, because even a dead jellyfish can sting you.

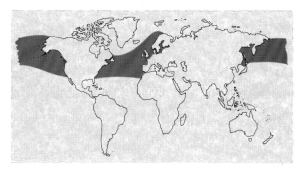

The lion's mane jellyfish grows to about 23 inches. ▼

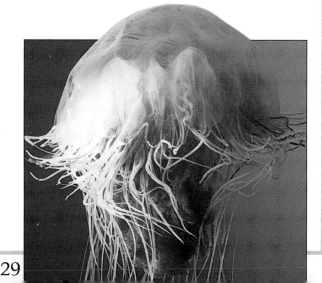

Glossary

Antennae Feelers on the head of an insect, lobster, or other animal.

Baleen Hard plates with bristles that strain food from the sea.

Blubber A thick layer of body fat in animals, such as whales and seals, that helps keep them warm.

Camouflage An animal's disguise to match the surrounding area.

Carrion A dead animal's body.

Cartilage A strong but flexible tissue that makes up the skeleton of sharks and rays.

Compound eyes Eyes containing light-sensitive cells that duplicate the same image many times.

Crustaceans A group of animals with armor-plated bodies, such as crabs and lobsters.

Echinoderm A group of spiny-skinned animals, such as sea urchins and starfish.

Echolocation The use of high-pitched sounds and their echo to move around or to detect food.

Evaporate To change from a liquid into a vapor.

Excretes To get rid of an unwanted substance from the body.

Extinction Having died out completely.

Food chains Where one animal is eaten by a larger animal, which is itself eaten by another larger animal. These animals are a chain of food for each other.

Lava Hot, molten rock that flows from volcanoes and becomes solid rock when cold.

Mangrove swamps A coastal habitat formed by the roots and branches of mangrove trees.

Mollusks A group of animals with soft bodies that are protected by shells (except octopuses and slugs).

Nocturnal Sleeping or resting during the day and active at night.

Nutrients Substances taken in by living things to help them grow.

Polar The climate of regions around the North or South poles.

Polyps Tiny coral animals that live in a colony.

Predators Animals that hunt and kill other animals for food.

Prey An animal that is hunted and killed by other animals.

Reptile A cold-blooded animal with scaly skin.

Sediments Material that settles at the bottom of the sea.

Spiracle An opening on the head of a ray fish, through which water can be pumped to the gills.

Temperate A climate that is warm, but not tropical or polar.

Tentacles Long organs, often near the mouth, for grasping food.

Tropical The climate of places that lie near the equator.

For Further Reading

Carwadine, Mark. <u>Whales, Dolphins, and Porpoises</u>. See and Explore Library. New York: Dorling Kindersley, 1992.

Collins, Elizabeth. <u>The Living Ocean</u>. Earth at Risk. New York: Chelsea House, 1994.

Hall, Howard. <u>Sharks: The Perfect Predators</u>. Close Up: A Focus on Nature. Morristown, NJ: Silver Burdett Press, 1994.

Lambert, David. <u>Seas and Oceans</u>. New View. Austin, TX: Raintree Steck-Vaughn, 1994.

Williams, Lawrence. <u>Oceans</u>. Last Frontiers. North Bellmore, NY: Marshall Cavendish, 1990.

Notes About Habitats

The world is divided into various habitat types, including deserts, grasslands, rain forests, temperate forests, mountains, and oceans. The distribution of these habitats is partly determined by the topography of the land and partly by the climate. Together, these two factors help shape the face of the planet. A way of classifying habitats is by the amount of rainfall they receive.

In some parts of the world, these different habitats have distinct borders, for example, when a forest meets the sea. However, it is more common for habitats to merge slowly into one another, such as a desert merging into a grassland. Consequently, some animals may be found in more than one habitat: Caracals may be found in deserts and grasslands, and birds of prey may soar over various habitats searching for food.

Index